LIVING PROCESSES

Plant Variation and Classification

Carol Ballard

WAYLAND

First published in 2015 by Wayland
Copyright © Wayland 2015

Wayland, an imprint of Hachette Children's Group
Part of Hodder & Stoughton
Carmelite House, 50 Victoria Embankment
London EC4Y 0DZ

Produced for Wayland by Calcium

Editors: Sarah Eason and Leon Gray
Editor for Wayland: Julia Adams
Designer: Paul Myerscough
Illustrator: Geoff Ward
Picture researcher: Maria Joannou
Consultant: Michael Scott OBE

Dewey number: 581.3-dc22
ISBN 978 0 7502 9655 7
10 9 8 7 6 5 4 3 2 1

MIX
Paper from
responsible sources
FSC® C104740
FSC
www.fsc.org

Printed in China

Every attempt has been made to clear copyright. Should there be any inadvertent omission please apply to the publisher for rectification. The author and publisher would like to thank the following for allowing their pictures to be reproduced in this publication:

Cover photograph: Shutterstock/leisuretime70.
Interior photographs: Dreamstime: Denis Baharew 19t; Fotolia: Jefery 35, Jrtb 31t, Evan Luthye 25b, Philippe Minisini 38b; Istockphoto: Ene 41tr, Keoni Mahelona 25t, Pedro José Pérez 18, Photoblaz 41tcl, Lord Runar 41bcl, Anna Yu 22; Shutterstock: Ajt 27t, asel101658 3, 5, 21, 30, Barbro Bergfeldt 43b, Adam Bies 26, Alex James Bramwell 10, 36b, 45, Chantal de Bruijne 19b, Sergey Chushkin 24, 37, Lawrence Cruciana 32, Pichugin Dmitry 29b, Dole 31b, Marie C. Fields 28, Roxana Gonzalez 40, Richard Griffin 43t, Sharon Harding 34t, Jubal Harshaw 9, 14, Hydromet 39, Iwka 34b, Javarman 7, Stepan Jezek 4, 27b, Kd2 36t, 38t, Anne Kitzman 39, LockStockBob 33b, Zacarias Pereira da Mata 20b, Maggie Molloy 43bl, Odze 11, Kamil Fazrin Rauf 13, Julie Royle 20t, Gordana Sermek 33t, James Thew 15, Vibrant Image Studio 6, Joao Virissimo 16, Surkov Vladimir 12; Wikimedia: 29tl, 29tc, 29tr, 41tl, 41tcr, 41bl, 41bcr, 41br.

An Hachette UK company   www.hachette.co.uk   www.hachettechildrens.co.uk

SAFETY NOTE: The activities in this book are intended for children. However, we recommend adult supervision at all times as neither the Publisher nor the author can be held responsible for any injury.

Contents

CHAPTER 1

What is classification?

Imagine a supermarket where cabbages are next to the tea bags and shampoo beside the tomatoes. It would be very hard to find the things you want. So supermarkets sort their products into categories such as fresh fruits and vegetables, tinned foods, pet foods and household cleaning items. This stops everybody getting muddled up.

In the same way as supermarkets sort and organise their products, biologists sort and group living things. It also helps people make sense of the living world, identify the right plant or animal, and understand the different relationships between living things. This process is called classification.

Common daisies are flowering plants found across Europe and the Americas. Their Latin name is *Bellis perennis*.

Carolus Linnaeus

A Swedish scientist named Carolus Linnaeus (1707–1778) created a way of naming living things based on the way they looked. His ideas form the basis of the classification system people use today.

Linnaeus gave every living thing a two-part Latin name. It works in a similar way to your own name. For example, a brother and sister might be called Mark Jones and Susan Jones.

Plants make their own food using the energy from the sun. The leaves are the food factories of green plants.

Jones is the family name. It shows that Mark and Susan belong to the same family. Each child also has his or her own first name – Mark or Susan.

Linnaeus used Latin names to identify living things. The Latin name for a human being is *Homo sapiens*. *Homo* is similar to the family name. The word *sapiens* is similar to the first name.

Latin names are always written in italics. Before the introduction of Linnaeus' naming system, scientists from different countries would use different names for the same living thing. As a result, no one could be sure whether they were talking about the same animal or plant. Latin was a universal language used by scientists in Europe, and so using Latin names solved this problem.

LIVING OR NOT?

Many things in the world are alive, but others are not. What are the differences between them? It all depends on whether or not something carries out basic life processes such as movement, feeding, growing and reproducing. Some non-living things can do some, but not all, of them. For example, a car uses fuel and it gets rid of waste. But a car cannot move by itself, or create new cars, so it is not alive. You might be able to think of some living things that do not seem to carry out all these life processes. Plants cannot move in the same way as animals, but this does not mean they cannot move at all. Sunflowers turn to face the sun, so just like humans, jellyfish, elephants, frogs and eagles, they are alive.

What makes a plant a plant?

Plants are living things. They are made up tiny units called cells and carry out life processes such as growing and reproducing. The ability to make their own food is common to all plants.

How do plants make their food?

Cells contain even smaller structures called chloroplasts. Chloroplasts contain a chemical called chlorophyll, which gives plants their green colour. In plants with different coloured leaves, some other pigments may also be present. Chlorophyll traps the energy in sunlight. Plants then use this trapped energy to convert carbon dioxide and water into food (glucose) and oxygen. The process is called photosynthesis.

This can be shown as a simple equation:

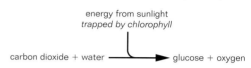

energy from sunlight
trapped by chlorophyll

carbon dioxide + water ⟶ glucose + oxygen

Plant cells

A simple animal cell has an outer layer called the cell membrane, which encloses a fluid called cytoplasm. The nucleus inside the cell contains genetic information in the form of DNA – deoxyribonucleic acid, the chemical that controls every feature of living organisms. Plant cells have a cell membrane, cytoplasm and nucleus, but they have other important structures, too. In the middle of a plant cell is a water-filled sac called a vacuole.

A network of branching tubes carries food and water from the leaves to different parts of the plant.

This keeps the cell rigid, preventing the plant from wilting. The plant cell has an extra outer layer, too, called the cell wall. This helps the cell keep its shape and acts as a barrier to stop chemicals entering and leaving the cell. In almost all plants, the cell wall is made from a substance called cellulose.

The stem of a tree is the woody trunk, shown here in cross section. Specialised cells in the stem form tubes that transport water from the tree's roots to its branches and leaves.

INVESTIGATE:
Specialised plant cells

Plants have special cells that do different jobs. Root hair cells help the plant absorb water from the soil. Palisade cells in the leaves carry out much of the photosynthesis for the plant. Look in books or on the Internet to find out more specialised plant cells. You could record your findings in a table like this:

Type of cell	Function
Guard cell	Open and close stomata, allowing carbon dioxide to enter leaves and allowing oxygen and water vapour to escape from leaves

CHAPTER 2

What is variation?

If you compare two living things, there will be lots of differences between them. This is called variation. For example, a fish has scales and fins, while a bird has feathers and wings.

Intraspecific or interspecific

A species is a group of living things that can breed and produce fertile offspring. Variation between two organisms of different species is called intraspecific variation. Variation between two individuals of the same species is called interspecific variation.

Characteristics

People are members of a species called *Homo sapiens*. If you look at people in your class, you will see many differences between them. Some will be tall, while others are short. Some will have brown, straight hair, while others have blonde, curly hair. Variations in characteristics like these can be used to distinguish between individuals.

The flowers of different plants come in many different colours, shapes and sizes. The beautiful red flowers of this hibiscus plant attract insects that help it reproduce.

Discrete and continuous variation

There are two types of variation – we call these discrete variation and continuous variation.

Discrete variation is an 'either/or' type of variation. The characteristic is either one thing or another and cannot be something in between. ABO blood type is an example of discrete variation – you can be either A, B, O, or AB.

Continuous variation refers to a characteristic that can have any value, within minimum and maximum limits. For example, a person could be just about any height, within limits. This is an example of continuous variation.

The leaves of the mint plant have serrated edges and are arranged in opposite pairs.

INVESTIGATE:
Variation in leaves

Look at the variation between leaves from a single plant. Choose a plant with plenty of leaves. Pick some leaves from the top, middle, bottom and each side of the plant. Before you start, decide on what your sample size will be. You have to strike a balance between measuring enough leaves to represent all of them, but not so many that your task becomes too time-consuming.

Measure the length of each leaf. Plot your results as a bar chart or use spreadsheet software to create a chart. Work out the average length of the leaves and the difference between longest and shortest leaves. Are any leaves exactly the average length? You could also measure the width of the leaves and compare them to the length. Is the ratio between length and width the same for all of the leaves?

Inherited variation

Some characteristics are passed on from parents to their offspring. They are called inherited characteristics. In people, eye colour, hair colour and ear shape are all types of inherited characteristics. Plants inherit characteristics, such as flower colour, in the same way.

Environmental variation

Some characteristics are not inherited. Instead, they are determined by the environment. For example, a plant will grow much taller and stronger with plenty of light, water and nutrients than a plant that does not get enough light, water and nutrients.

Most plants need water to carry out their basic processes. If they are deprived of water for a long period of time, such as during a severe drought, many plants will die.

Inheritance and environment

Some inherited characteristics are also affected by the environment. For example, you may have inherited the ability to be an excellent musician, but if you never hear music you will not develop your talents.

Variation, natural selection and evolution

In the nineteenth century, the English scientist Charles Darwin (1809–1882) came up with a new theory called 'natural selection'. He said that some individuals within a species might be better suited to their environment than others. For example, a plant with strong roots would survive better in a drought than a weak-rooted plant of the same species. In a dry region, the plant with strong roots would survive and pass on this characteristic to its offspring.

Over many generations, the species would change, or evolve. Natural selection ensures that the fittest individuals survive and reproduce while the weaker individuals do not. This process of change is called evolution.

Evolution is responsible for the wide range of different plant species found in a tropical rainforest.

MUTATIONS

Mutations are changes that happen when cells divide. Some mutations have no noticeable effect. Some mutations are beneficial to the organism, but others are harmful. Some mutations arise naturally just as a result of errors made during cell division, while others are caused by exposure to radiation or chemicals. If a plant with a mutation reproduces, the characteristics of its offspring may be different from those of the parent plant. If the changes are beneficial, natural selection will lead to plants with the mutation increasing in number. If the changes are not beneficial, the mutated plant is unlikely to survive and reproduce.

CHAPTER 3

How to classify

Scientists have identified many different species since Linnaeus was alive and have changed his system to include them all. Today, most scientists group living things into five separate groups called kingdoms. The five kingdoms are: bacteria, protists, fungi, plants and animals.

DNA analysis

Scientists can now compare the DNA of different plants. Using this information, they can work out the relationships between them and how new types have evolved. DNA analysis has shown that some classifications are incorrect, and new groupings are necessary. This means that plant classification is constantly being changed and reorganised.

The plant kingdom

There are more than 250,000 different plants in the plant kingdom. Features that are common to all plants include the ability to make their own food and the presence of a cell wall around the cells that make up the body of the plant.

A cross section through the stem of a cotton plant reveals bundles of cells (green). These make up the tubes that carry water through the plant. The cotton plant belongs to the group of vascular plants.

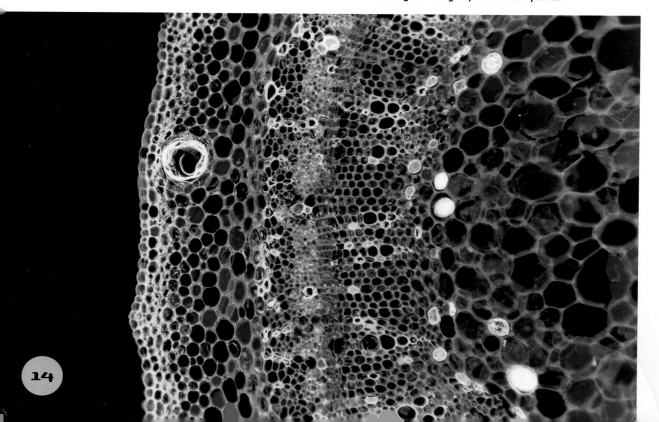

Dividing the plant kingdom

The plant kingdom is divided into smaller and smaller groups as follows:

Plant kingdom
Division
Class
Order
Family
Genus
Species

The plant kingdom is subdivided into plants that do not have a water-transport (vascular) system and those that do. Some scientists think that non-vascular plants are not true plants and should not be in the plant kingdom at all. Vascular plants are split into two groups – seed-bearing plants and those that do not have seeds. Each separate group splits again into smaller and smaller groups.

Dandelions are seed-bearing plants. The seeds are blown away in the wind. If they land in a suitable spot, the seeds will grow into new dandelions.

OTHER CLASSIFICATION SYSTEMS

The Greek philosopher Aristotle (384–322 BCE) classified living things as either plants or animals. His system lasted until the nineteenth century, when scientists added a third kingdom of microorganisms, which included bacteria. An American scientist named Robert Whittaker (1920–1980) came up with the modern five-kingdom system in 1969. From recent studies, including DNA analysis, some scientists think that a six-kingdom or even a seven-kingdom system would be better still.

Identifying plants

Plants are identified and classified by examining some of their main features. To do this, scientists use a key. There are two main types of keys – branching keys and numbered keys. Each key works in a similar way. It asks a series of questions that lead scientists to the name of the plant they wish to identify.

Which characteristics?

There is a wide range of features that can be used to help identify a plant. Some include plant parts such as leaves, fruits, seeds, stems and bark. Others relate to the plant as a whole, such as the shape of a tree or how close the plant grows to the ground. Still others relate to the plant in its environment, such as the type of habitat in which it grows or the season during which the plant produces flowers.

These cacti survive in the scorching heat by storing water in their huge stems.

16

INVESTIGATE:
Using keys

Numbered keys begin with a question, for example, 'Is the nut rough?' If you answer yes, you go to one question. If you answer no, you go to another question. Eventually, you end up at the plant you are trying to identify. Branching keys also begin with a question. The branching key will have a branch to follow if the answer is 'yes' and another branch if the answer is 'no'.

Identify each seaweed using these keys.

Numbered key

1	Is it green?	yes	go to 2
		no	go to 3
2	Does it look like blades of grass?	yes	eel grass
		no	sea lettuce
3	Is it brown?	yes	go to 4
		no	go to 6
4	Are it edges toothed?	yes	toothed wrack
		no	go to 5
5	Does it have swollen lumpy parts?	yes	bladder wrack
		no	kelp
6	Does it branch regularly into two?	yes	carrageen
		no	dulse

Branching key

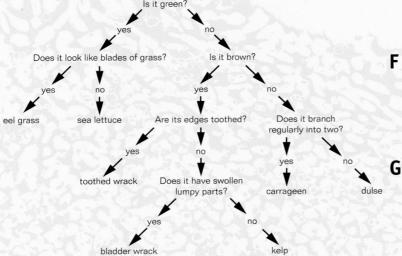

A

B

C

D

E

F

G

CHAPTER 4

Simple plants

In the past, scientists classified algae as simple plants. Like all plants, algae can make their own food by photosynthesis, but they do not have a vascular system. Some scientists think that algae do not belong to the plant kingdom. Instead, they suggest that algae are protists. In this book, however, algae are treated as simple plants.

Common characteristics

Algae do not have separate roots, stems and leaves. Instead, they consist of a single structure, called a thallus. Algae live in water or in damp places and can absorb water through every part of their surface.

Algae reproduce in a two-stage process. First, male and female sex cells join to form an embryo. The embryo then develops into an alga that produces spores. The spores are released and germinate to form new plants.

Many algae live on the surface of ponds. Some algae grow so quickly that they clog up the water and suffocate other plants and animals such as fish.

Seaweeds

The most familiar algae are seaweeds, which are found on rocks, in rock pools and floating up from the sea bed along most shores. There are three main types of seaweed, classified by their colour – red, green and brown. Each has different coloured pigments, but all can make their own food by photosynthesis.

Other algae grow on stones and in damp places around ponds and other freshwater habitats. Some are small and delicate, while others are large and strong.

The leaf-like lamina of this green seaweed make food for the plant by the process of photosynthesis.

NOT QUITE A PLANT?

When scientists first started to classify living things, they grouped them as either animals or plants. Many organisms were classified as plants, but scientists now think they belong elsewhere. These organisms include:

- Fungi such as mushrooms, toadstools and moulds. Fungi grow on rotting fruit and share many characteristics of plants, but they cannot produce their own food.
- Blue-green algae and bacteria. These single-celled organisms do not contain nuclei like plant or animal cells. Most are parasites that steal nutrients from other living things.
- Protists such as amoebas. These single-celled organisms live in water and are too small to be seen without a microscope.

Some protists, such as *Euglena* and *Chlamydomonas,* show many plant-like characteristics, such as the ability to make their own food, while others are more like animals and feed on other microorganisms.

Fungi, bacteria and protists are all classified in separate kingdoms.

Mosses and liverworts

Mosses and liverworts have more complex structures than algae. Most grow in clumps in damp, shady places. They are found in almost every part of the world, including deserts and polar regions. There are more than 6,000 different species of mosses and 8,000 species of liverworts.

Common characteristics

Mosses and liverworts are non-vascular plants. Although they do not have roots, they do contain fine, root-like structures called rhizoids. These absorb water and nutrients and also help to anchor the plant in its place. Most mosses have an upright shoot, with spirals of tiny leaf-like structures. Most liverworts have a thin, flat body called a thallus that grows flat on moist soil or on the surface of still water.

Liverworts thrive on land or water in dark places where there is little sunlight.

Mosses cover rocks that litter the forest floor. Fine, root-like structures called rhizoids anchor the moss to the rocks.

Reproduction

Like algae, mosses and liverworts reproduce in a two-stage process. In the first stage, male and female sex cells join up to form structures called sporophytes. The sporophytes produce tiny spores in small capsules held above the plant on thin stalks. The plant then releases the spores, which are dispersed by the wind. Eventually, the spores germinate and grow into new mosses or liverworts.

Common examples

Funaria hygrometrica is a common moss found in many climate zones around the world. The stalk that holds its spore capsule is a spiral. In dry conditions, the stalk curls up. In damp conditions – which are ideal for germination – the stalk unwinds.

The common liverwort *Marchantia polymorpha* is found growing on the soil of many pot plants. It needs wet conditions, and its natural habitat is close to rivers and other freshwater sites.

BOG MOSSES

There are between 150 and 350 species of *Sphagnum* moss. Many are found in peat bogs and are sometimes called bog mosses. *Sphagnum* plants can store both water and air well; some can hold as much as 20 times their dry weight of water! In dry conditions, spaces inside the plants act as water-storage containers. In wetter conditions, the spaces fill with air and help the moss to float. Because of its water storage properties, *Sphagnum* moss is often used to line hanging baskets and other plant containers.

Bog moss is the common name given to plant species from the genus *Spaghnum*. These plants live in peat bogs around the world.

CHAPTER 5

Ferns, horsetails and club mosses

Ferns, horsetails and club mosses are simple plants that live on the land. Recent DNA studies suggest that they may not be as closely related as scientists first thought.

Ferns

There are more than 10,000 different fern species. Most grow in tropical regions, but many are also found in temperate regions, some live in deserts and a few live in the tundra. Ferns come in many different shapes and sizes, but they all share certain features. The leaves of a fern are called fronds. They have a long central part, called the rachis, from which many small leaflets spread out. At first, the fronds are tightly coiled but uncoil as they mature.

The largest ferns are tree ferns such as *Dicksonia antarctica*, which can grow up to 20 metres tall, with fronds more than 5 metres long. The central stem of a tree fern looks like a tree trunk. However, it is not a woody trunk but a stem-like structure called a rhizome that supports the plant. The fronds emerge from the tip of the rhizome.

Fern reproduction

Ferns reproduce using tiny structures called spores. On the underside of the fronds are clusters of small, brownish structures called sporangia. These produce spores. After their release and dispersal, each spore germinates to form a new fern plant.

Azolla **ferns form a dense carpet over the surface of a pond, crowding out any other aquatic plants.**

INVESTIGATE:
Using keys

Use this simple numbered key to identify these ferns.

1. Are its fronds divided?

    yes   go to 2
    no   Hart's tongue fern

2. Are the fronds patterned with grey and silver?

    yes   Japanese painted fern
    no   go to 3

3. Are the pinnules forked?

    yes   Fishtail fern
    no   Bracken

A

B

C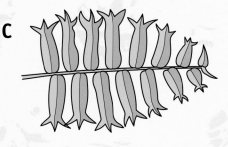

D

Helpful or harmful?

The smallest ferns, such as *Azolla*, live in the water. They float on the surface of lakes and ponds in tropical and subtropical regions. The tiny fronds of *Azolla* are just 1 centimetre long. *Azolla* plants contain a microorganism that absorbs a gas called nitrogen from the air. This helps *Azolla* grow rapidly. The plant can double in size every few days. In many parts of the world, *Azolla* is a weed. It covers large areas of fresh water and disrupts the natural ecology of the habitat. In China, *Azolla* plants are useful to rice farmers. The farmers grow *Azolla* in rice paddies in the spring. The *Azolla* grow quickly to cover the water, preventing weeds from growing in it. When the *Azolla* plants die, they release nitrogen back into the water. The rice plants then use the nitrogen to grow into healthy crops.

Horsetails

There are fewer than 30 horsetail species. They grow in most temperate and tropical parts of the world, but they are not found in Australia and New Zealand. They are simple vascular plants, with whorls of small, scale-like leaves around hollow, jointed stems that grow from underground rhizomes.

Horsetails reproduce using spores, which are produced in cone-like structures at the tips of the stems. Some, such as a tropical American species, *Equisetum giganteum*, can grow up to 5 metres high, but most are only about 20 centimetres tall. The horsetail *Equisetum arvense* is a weed hated by many gardeners. Like other horsetails, the rhizomes spread through the soil, so even when one piece is pulled up, others quickly appear close by.

Club mosses

There are about 1,000 different club mosses, most of which grow in moist areas in tropical and subtropical forests. Club mosses are like mosses, but they have a vascular system. They have true roots, a stem and tiny, scale-like leaves. Club mosses reproduce using spores, which are either produced in small cones or at the junctions between leaves and stem.

The wood horsetail lives in forests across the northern hemisphere. The rhizomes of the plant spread through the soil, and new plants spring up through the forest floor.

Club mosses have leaves that look like those of a conifer.

Stag's horn club moss *Lycopodium clavatum* is the most common club moss. Its tiny yellow spores form a powder that easily catches fire. For this reason, it has been used to make fireworks.

A club moss called *Selaginella lepidophyllum* grows in the deserts of south-west United States and Mexico. It survives in the desert by curling into a brown ball. When it rains, the moss uncurls and turns green. It is also called the 'resurrection plant' because it seems to come back from the dead.

SCOURING AND SANDING

Horsetails are also called scouring rushes. They get this name because they look a bit like plants called rushes, and the stems, right, contain a rough abrasive substance called silica. In many countries, horsetails are traditionally used to make scouring pads. In Japan, the horsetail *Equisetum hyemale* is used as sandpaper to smooth and polish wood.

CHAPTER 6

Plants with cones

There are four groups of plants that have seeds but no flowers. These are the conifers, cycads, gingko and the gnetophytes. Together, they are called gymnosperms, which comes from the Greek meaning 'naked seed'. The seeds of gymnosperms are 'naked' because they are not contained in an ovary as they are in plants with flowers.

Cone characteristics

There are between 700 and 800 gymnosperms found in a wide variety of habitats. Most are long-lived trees or bushes with woody stems called trunks. They have specialised roots and leaves and a well-developed vascular system.

Gymnosperms produce cones that contain either the male or female reproductive structures. Most cones are hard, but in some conifers they are soft and fleshy. Male cones produce pollen grains, which are carried to the female cones by wind. Eventually, the pollen

Conifer trees form vast forests in northern temperate and mountainous regions across North America, Europe and Asia.

The red berry-like structures are the mature arils of the European yew. The green arils are not fully mature.

joins with the egg. Seeds develop, which are dispersed by the wind. They germinate and develop into new plants.

Common conifers

Conifers are the common pine trees or fir trees. Most are evergreen, with scale-like or needle-like leaves. The word conifer means 'cone bearing' – all conifers bear cones. Some common conifers include:

- Yew trees (*Taxus baccata*) are unusual conifers because they do not have normal cones. Instead, their seeds form inside round (usually red), fleshy structures called arils.
- Giant redwoods (*Sequoiadendron giganteum*) are the largest trees on the planet. They can live for more than 2,000 years. Some are more than 110 metres tall, with trunks nearly 8 metres in diameter.
- Monkey puzzle trees (*Araucaria araucana*) originate from Chile. They get their name because people thought that monkeys would not be able to climb up their scaly branches.

WEATHER FORECASTING

Pine cones have been used to predict the weather. In dry conditions, pine cones lose moisture into the air. The scales of the cones become stiff and shrivelled. In damp conditions, the cones absorb moisture from the air. The scales become flexible and return to their normal shape.

Cycads

It would be easy to mistake a cycad for a palm tree, because they look very similar. Cycads have cones, however, and palms do not. Cycads are found in tropical and subtropical parts of the world. Some grow up to 15 meters tall, but most are much shorter. The crown of leaves forms a rosette at the top of the trunk. Male and female cones are produced by separate plants. The female cones are large – some weigh more than 30 kilograms – and are supported among the leaves.

The leaves of *Gingko biloba* are similar to the pinnae of the maidenhair fern.

Gingko

Gingkos are an ancient group of plants, but only one species survives today – the maidenhair tree (*Gingko biloba*). This gets its common name because the fan-shaped leaves resemble the leaf-like pinnae of the maidenhair fern. These large trees grow up to 30 metres tall and can live for thousands of years. Unlike most other gymnosperms, the maidenhair tree has delicate, fan-shaped leaves. Its soft, fleshy seeds have an unpleasant smell. This species is now very rare and can only be found in the wild in a small area of eastern China.

Gnetophytes

Gnetophytes are the closest surviving ancestors of the flowering plants. There are three groups: *Gnetum*, *Ephedra* and *Welwitschia*.

Gnetum species are tropical woody climbers and trees. They have simple broad leaves, and their seeds are traditionally used to make a crispy snack.

Ephedra, or joint-fir, is found in deserts throughout the world. It is a bushy shrub with green, jointed stems and small scale-like leaves.

Welwitschia lives in the Namib Desert of south-western Africa. There is just one species, *Welwitschia mirabilis*. The trunk is mostly buried, with two strap leaves attached to it.

These leaves grow continuously and live as long as the plant itself. They slowly tear into long strips. Male and female cones are produced on the trunks of different plants.

INVESTIGATE:
Using keys

Use this simple numbered key to identify these gnetophytes.

A  **B**  **C**

1	Large, strap-like leaves?	yes	Welwitschia	2	Simple broad leaves?	yes	Gnetum
		no	go to 2			no	Ephedra

Welwitschia mirabilis is a desert-dwelling gnetophyte that may live for up to 2,000 years.

CHAPTER 7

Flowering plants

There are about 240,000 different flowering plants. They come in many different shapes and sizes and survive in a range of climates and habitats. Some flowering plants live for just a few months, some for a year and others for hundreds of years. Oak trees, daffodils, grasses, water lilies, stinging nettles and carrots are all flowering plants.

An oak tree bears fruits called acorns. Each acorn contains one seed, which is dispersed by animals who eat the acorns, such as squirrels.

Flowering plants are often called angiosperms. The word angiosperm comes from a Greek word meaning 'enclosed seed'. The seeds of angiosperms are 'enclosed' because they are kept inside an ovary.

Flower features

All flowering plants share certain common features. For example, they have fully developed vascular system, roots, stems and leaves. And of course, they all produce flowers. The flowers contain the plant's reproductive parts. Some flowers contain male parts called stamens. Other flowers contain female parts called the stigma, style and ovary. Still others have both male and female structures. It is from these reproductive structures that seeds develop, which can then grow into new flowering plants.

Roots absorb water and nutrients from the soil and anchor the plant in the ground. Some plants have long roots that grow straight down into the soil, while others have roots that form fine, branching networks near the surface. Spreading roots produce shoots some distance from the parent plant. These shoots can develop into new plants. In other plants, the roots swell up and act as food stores.

The leaves and flowers of the water lily float on the surface of a pond. The roots are submerged and anchor the plant in the mud.

Stems can be just about any length and thickness. Many are soft and green, while others are strong, hard and woody. Stems carry water from the roots to the rest of the plant. They also support the leaves and flowers.

Leaves also come in many shapes and sizes. Some are very thin, while others have a tough, waxy covering. The leaves contain chloroplasts, which trap energy from sunlight and use it to make food in the process called photosynthesis.

FLOWERING TIMES

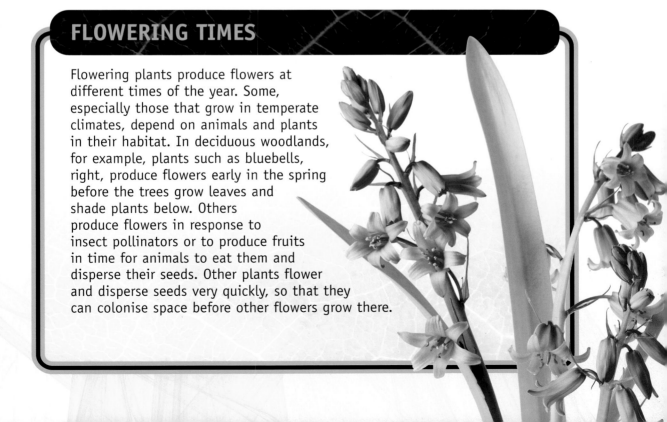

Flowering plants produce flowers at different times of the year. Some, especially those that grow in temperate climates, depend on animals and plants in their habitat. In deciduous woodlands, for example, plants such as bluebells, right, produce flowers early in the spring before the trees grow leaves and shade plants below. Others produce flowers in response to insect pollinators or to produce fruits in time for animals to eat them and disperse their seeds. Other plants flower and disperse seeds very quickly, so that they can colonise space before other flowers grow there.

Flowers

Flowering plants grow from seeds. When a plant is mature, it produces flowers. Flowers share some basic features. Leaf-like sepals protect the flower in the bud and form the flower base when the petals open. The reproductive parts are in the centre of the flower. The male part of the flower, called the stamen, produces pollen grains. The female reproductive parts are the ovary, which contains one or more ovules with egg cells, and the stigma, which is held up by a stalk called the style.

THE REPRODUCTIVE ORGANS OF A FLOWER

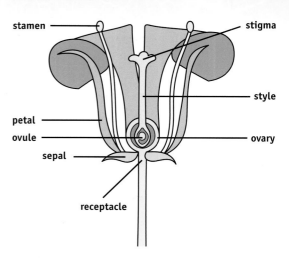

stamen · stigma · style · petal · ovule · ovary · sepal · receptacle

Pollination

Before seeds can develop, male pollen must land on the female stigma. This process is called pollination. In some plants, pollen from one flower can pollinate another flower on the same plant. This is called self-pollination. In other plants, the female parts of one plant receive pollen from another plant.

In plants such as grasses, pollen is carried by the wind. Other plants rely on insects such as bees, butterflies and moths to carry the pollen from flower to flower. That's why many flowers are brightly coloured – the colour attracts pollinators. Others produce sweet nectar to attract insect pollinators, while some produce scents to attract moths at night.

The colourful, fragrant flowers in this field attract pollinators such as bees and other insects.

Fertilisation

When a flower has been pollinated, a pollen tube grows from the pollen grain down through the style. A pollen cell travels down the pollen tube and fertilises the egg inside the ovule.

Fruits and seeds

Each fertilised egg will develop into a seed. This contains the embryo, which will give rise to a new plant when it germinates, and a food store for the embryo until it can produce its own food by photosynthesis. The embryo and food store are enclosed in a seed coat or testa.

The primrose is a colourful flowering plant, shown here with blue and yellow flowers. Gardeners have developed a range of flower colours, including purple, red, pink and yellow.

INVESTIGATE:
Germination

Germination is affected by many factors. Investigate them for yourself by growing your own seeds. Buy some cress seeds from your local garden centre. Lay a piece of kitchen towel on four saucers. Sprinkle 20 small seeds onto each piece of kitchen towel. Pour water to cover the seeds on three saucers. Put one in a cold, light place. Put one in a warm, light place. Cover one with dark card. Put the dry saucer in a warm, light place. Check your seeds every day for one week and record how many seeds on each dish have germinated. Under which conditions did the seeds germinate soonest? Did any germinate without water? What do you think is most important for germination: light, warmth or water?

Life cycles of flowering plants

All flowering plants reproduce using seeds, but their life cycles may be very different. Annual plants germinate, flower and produce seeds in one year. The parents die and are replaced by new plants when the seeds germinate the following year. Sunflowers, wild poppies and forget-me-nots are annuals.

Biennial plants germinate and grow in the first year of life. In the second year, they flower and produce seeds. The parents die and are replaced by new plants when the seeds germinate the following year. Wallflowers and hollyhocks are biennials.

Forget-me-nots develop from tiny seeds that grow in pods on the stem. The pods stick to animal fur and clothing and are carried away from the parent. The pods eventually fall off, and the seeds inside may grow into new plants.

The flowers of the common hollyhock range in colour from dark red to pink, yellow and orange.

Perennial plants live for many years. After germination, some perennials flower and produce seeds in their first year. Others take many years to mature. Maple trees, roses, primroses and honeysuckles are perennials.

Some perennial plants, such as peonies and hostas, die back to their roots every autumn, leaving dead stalks above the ground. In spring, new shoots grow from which the new season's plants develop.

Deciduous or evergreen?

Many perennials lose their leaves every autumn. The branches remain bare through the winter. New leaves grow the following spring.

The huge flowers of the rafflesia plant burst open and wilt away within the space of one week.

Plants that do this, such as oak trees and rose bushes, are called deciduous plants. Other perennials keep their leaves all year round. They are called evergreen plants. Holly and laurel are evergreens.

Cuttings and clones

Flowering plants produce seeds, which grow into new plants. New plants often grow from spreading roots. Gardeners and horticulturists also grow new plants from cuttings and by cloning plant tissues. Plants that grow from spreading roots or from cuttings or cloning are genetically identical to the parent plant.

PLANT EXTREMES

The biggest flowers belong to the aptly named monster flower, *Rafflesia arnoldii*. This plant grows in the rainforests of Indonesia and Malaysia. Each flower can measure more than 1 metre across and weigh 10 kilograms. The flowers smell like rotting flesh, which attracts the flies that pollinate them.

At the opposite end of the scale are the flowers of *Wolffia arrhiza,* a type of duckweed that grows in Europe, Asia and the Americas. The tiny plants, which float on water, are between 1 and 5 millimetres long. These plants rarely produce flowers, but when they do each flower is less than 0.3 millimetre across.

Seed dispersal

If seeds germinate near to the parent plant, they will be competing for resources such as space, light, water and nutrients. This will stop any of them from growing into strong, healthy new plants. To avoid this, the seeds must spread as far away from the parent plant as possible. This is called seed dispersal. The structure of the seed and/or fruit can often indicate how the seeds are dispersed. Some seeds are enclosed inside fruits. Fruits such as nuts and pea pods are dry. Other fruits, such as strawberries and tomatoes, are soft and fleshy.

A butterfly feeds on the sugary nectar of a flower. In return, the butterfly pollinates the flower.

Wind and water

Some seeds are dispersed in the wind. For example, the 'helicopter' seeds of sycamore trees and the 'clocks' of dandelions are light and 'fly' in the wind. Other seeds are produced inside cases that have tiny holes in them. As the plant stem moves in the wind, the ripe seeds fall out through the holes. Poppies disperse their seeds in this way. Some water plants, such as pond irises, produce seeds that can float, so they are dispersed by water currents. Others produce seeds with woody, waterproof coverings that survive in water for long periods. One example is the coconut, which can be carried thousands of kilometres across the ocean.

The coconut is the fruit of the coconut palm. It is light and floats in the water, which aids its dispersal in the ocean currents.

Using animals

Brightly coloured, fleshy fruit attracts animals and birds. They eat the fruit and the seeds they contain. The seeds pass through the animal's body intact. They pass out of the animal's body in its waste, often some distance from the parent plant. Blackberries are dispersed in this way.

Other plants, such as burdock, produce seed heads with hooks or barbs that stick to fur as an animal brushes past the plant. The seed head, with its seeds inside, is carried away by the animal, eventually falling off elsewhere.

EXPLOSIONS!

Some plant seeds, such as those of gorse bushes, are enclosed in seed cases that dry and ripen with the seeds. When the seeds are fully ripe, the seed case explodes and the seeds shoot out into the air.

The prickly seed heads of the burdock plant inspired Swiss engineer George de Mestral to invent the hook-and-loop fastener called Velcro.

Dicotyledons and monocotyledons

The food store inside the seed of a flowering plant is called a cotyledon. There are two groups of flowering plants with different number of cotyledons in their seeds. If the seeds of a plant have one cotyledon, it is called a monocotyledon, or monocot for short. If the seeds have two cotyledons, the plant is know as a dicotyledon, or dicot.

Bamboo is the fastest-growing plant on Earth. Some species form dense forests in tropical regions around the world.

Monocots

Scientists have identified around 65,000 monocot species. Grasses are some of the most important monocots. Grasses include crops such as barley, maize, oats, rice and wheat, as well as bamboo and sugar cane. Many monocots are small plants, but some, such as palms and pineapples, grow much larger.

As well as the single cotyledon, monocots share some common features. Their leaves are usually long and strappy, with veins that run straight from base to tip without branching. If you look at a cross section of a monocot stem under a microscope, you will see bundles of vascular tissue scattered randomly throughout the stem. Few monocots have woody stems.

Wheat is an important cereal crop. The grain is ground to make flour for bread and foods such as pasta and noodles.

CEREALS

Cereals are an important source of food for people. They provide energy and essential nutrients. Different cereals are grow in different parts of the world. This is because different plants are suited to different climates. This table shows some of the most important cereals eaten by people from different parts of the world.

Cereal	Important part of diet in
maize	North America, Africa, Australia, New Zealand
rice	many tropical regions
wheat	all temperate regions
sorghum	Asia and Africa
rye	colder regions
millet	Asia and Africa
quinoa	South America

Monocots produce flowers with parts in multiples of three. For example, irises have three sepals, three outer petals and three shorter inner petals. Although they can be many different shapes, the pollen grains of monocots usually have just one pore or furrow. The seeds are often contained in a three-sectioned capsule.

The yellow kernels of sweet corn are the fruits of a type of maize. The kernels are held on a cob and wrapped in leaves called the husk.

Bulbs

Many monocots produce bulbs, which are modified leaves. These serve as food stores for the plant over the winter. New roots and shoots develop from the bulbs in the spring. Onions and garlic both grow from bulbs, as do lilies and tulips.

Dicotyledons

Scientists have identified about 200,000 different dicots, ranging from huge, long-lived trees to tiny, short-lived annuals. They are found in most habitats on Earth. Dicots are vitally important to people. They provide us with fruits and vegetables, crops such as tea and coffee, many medicines, and products such as cotton, rubber and timber.

As well as the two cotyledons, dicots share some common features, which can be used to tell them apart from monocots. The leaves come in a wide range of shapes and sizes and usually have a network of branching veins, often running from a larger central vein. If you look at a cross section of a dicot stem under a microscope, you will see bundles of vascular tissue arranged in a ring. Many dicots have woody stems.

Dicots usually produce flowers with parts in fours or fives. For example, primroses have five sepals, five petals joined in a tube and five stamens. Some produce flowers with parts in multiples of four or five. For example, Mountain avens gets its Latin name *Dryas octopetala* because it has eight petals. The word *octo* means 'eight' in Greek. Like the monocots, the pollen grains of dicots can be many different shapes but usually have three pores or furrows. The seeds are often contained in four- or five-sectioned capsules.

New classifications

Flowering plants have traditionally been split into monocots and dicots. Some botanists now think that the dicots themselves should not be classified as a single group. They think that around 70 per cent of the dicots should remain in the dicot group, while the rest should be split up into smaller groups, such as magnolias and water lilies. Scientists are still debating this issue, though, so the old system is still widely accepted.

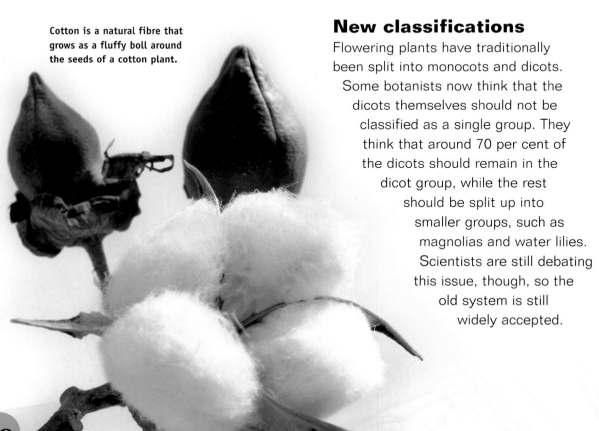

Cotton is a natural fibre that grows as a fluffy boll around the seeds of a cotton plant.

INVESTIGATE:
Using keys

Use this simple branching key to identify these flowering plants.

A   B   C   D

E   F   G   H

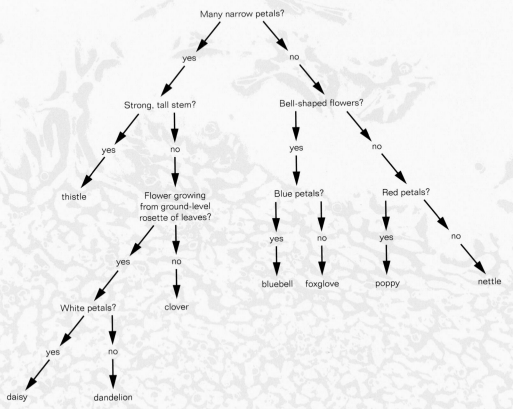

Many narrow petals?

yes → Strong, tall stem?

no → Bell-shaped flowers?

Strong, tall stem?
- yes → thistle
- no → Flower growing from ground-level rosette of leaves?

Flower growing from ground-level rosette of leaves?
- yes → White petals?
- no → clover

White petals?
- yes → daisy
- no → dandelion

Bell-shaped flowers?
- yes → Blue petals?
- no → Red petals?

Blue petals?
- yes → bluebell
- no → foxglove

Red petals?
- yes → poppy
- no → nettle

Create a key

In this activity, you will create your own branching key and use it to identify leaves collected from different plants. Remember it will be much easier to highlight differences between leaves if you look at a range of different plants. This will make it easier to create your key. Try to select leaves of different sizes, shapes, colours and textures.

Look at the trees that line the roads and in your local park. You might find a broad-leaf tree such as an oak or sycamore or a conifer such as a pine.

Smaller garden plants and bushes would also provide leaves of different shapes and sizes. And you could even take leaves from plants around your home, too.

If you have access to a camera, take a picture of the plant you have chosen for your leaf sample. This will make it easier to identify the plant if you have trouble finding it. The guide below highlights some of the features you can use to identify the leaves of different plants.

IDENTIFYING BY THEIR LEAVES

Sycamore
Descriptive: the sycamore tree has simple, large leaves arranged in opposite pairs around the twigs. There are five distinct lobes to each leaf, and five veins spread out from the base of the leaf into the lobes. The leaves have ragged edges, with lots of rounded 'teeth'. There are some hairs on the lower surface of the leaf.

Stinging nettle
Descriptive: the stinging nettle has dark green oval-shaped leaves, with long stalks. The tip of the leaf is pointed, and the base is heart-shaped. The leaves are several centimetres long and the edges have lots of coarse 'teeth'. The leaves are covered with tiny hairs that result in a painful sting when touched. REMEMBER TO WEAR YOUR GLOVES WHEN COLLECTING LEAVES FOR THIS ACTIVITY.

Swiss cheese
Descriptive: the split leaf *Philodendron* or Swiss cheese plant is a common house plant. It has perforated broad leaves that are deeply split along the edges. The dark green leaves can be enormous for a house plant, sometimes up to 1 metre across and they have a waxy texture.

You will need:

- paper
- card
- sticky tape
- writing materials
- a selection of different leaves
- some reference books and/or access to the Internet

1. Collect some leaves from different plants in your garden or local park. Take the leaves home.

2. Examine the leaves. Can you spot any features that you could use to split them into two roughly equal groups? For example, you could divide them into leaves with simple shapes and those divided into lobes or leaflets. Or the leaves but be narrow or wide.

3. Sort the leaves into the two groups that you have chosen. At the top of your paper, in the centre, write the first question you used to split your leaves into two groups. (For example: 'Are the leaves narrow?') Draw two arrows diagonally down to the left and right. Write 'Yes' on the left arrow and 'No' on the right arrow.

4. Now look at your 'Yes' group. Can you find another feature that you could use to split these leaves into two groups? Write the question and then draw two more 'Yes' and 'No' arrows.

5. Repeat step 4 for your 'No' group.

6. Repeat these steps until each leaf is in a single group on its own.

7. Use reference books or the Internet to identify the first leaf. Write the name below the final question in the key.

8. Repeat step 7 until you have identified all the leaves and written all their names in the key.

9. Go back to the beginning of the key. Pick a leaf at random and follow it through the key. If your key works, you should reach the correct name of the leaf. Try it for some of your other leaves.

10. Mix your leaves up then stick them on a sheet of card in and number them 1, 2, 3, etc.

11. Give a friend your leaves and key. Can they identify each leaf correctly?

Glossary

adaptation Characteristic that helps an organism to survive in its surroundings.

algae Plants that do not have leaves, stems, roots or flowers.

angiosperm Plant that produces seeds and flowers.

annuals Plants that grow from seed, reproduce and die in one season or year.

bulb Modified leaves that serve as a food store and from which a new plant can grow.

cell The basic building block of all living organisms.

cell membrane Thin 'skin' enclosing the cytoplasm and organelles in all cells.

cellulose Fibrous material from which most cell walls are made.

cell wall Strong outer layer of a plant cell.

chlorophyll Green pigment found in the cells of green plants.

chloroplast Organelle in green plant cells that contains chlorophyll. Chloroplasts are the site of photosynthesis.

chromosome Thread-like structure that carries genetic information.

conifer Tree adapted to colder climates with needle-like leaves that cover the branches throughout the year.

class Classification group that contains one or more orders.

cytoplasm Watery fluid inside a cell.

deciduous Trees that lose their leaves and lie dormant once a year.

dicotyledon Plant that produces seeds with two food stores called cotyledons.

division Classification grouping that contains one or more classes.

DNA Deoxyribonucleic acid, the chemical from which chromosomes are made.

embryo Organism in its early stages of growth.

evolution Process by which every living thing slowly changes over time because of the slight variations in genes from one generation to the next.

family Classification grouping that contains one or more genera.

fertilisation Fusion of pollen cell and egg cell in a flowering plant.

fronds Leaf-like parts of ferns.

fungi Plant-like organisms that do not contain chlorophyll and so cannot make their own food. Mushrooms, moulds and yeasts are fungi.

gene Sequence of DNA that carries one piece of genetic information.

genus Classification grouping that contains one or more species.

germination The beginning of growth of a seed or spore.

gymnosperm Plant that produces seeds but not flowers.

interspecific Between two species.

intraspecific Within a single species.

meiosis Type of cell division that produces four non-identical sex cells.

mitosis Type of cell division that produces two identical cells.

monocotyledon Plant that produces seeds with one food store called a cotyledon.

mutation Change in the genome of an organism.

nectar Sweet liquid produced by the flowers of some plants.

non-vascular Lacking a water-transport system.

nucleus Structure inside a cell that contains genetic information.

nutrients Substances that organisms need to live.

order Classification grouping that contains contains one or more families.

parasite Organism that lives on a host organism. The parasite uses the host as a source of food or for protection and may harm, or even kill, it.

pollen Grains that contain the male reproductive cells of a seed plant.

pollinate Transfer pollen from the flower of one plant to another plant of the same species.

rhizoid Root-like structure of some simple plants.

rhizome Stem-like structure that spreads underground in some plants.

sap Juice inside plant stems and leaves.

seed Reproductive unit of flowering and some non-flowering plants.

species A group of organisms that can mate with one another and produce fertile offspring.

sporangium Structure that produces spores in ferns.

spore Reproductive unit of some plants and fungi.

thallus Main part of simple plants such as liverworts.

vacuole Space within a plant cell that contains cell sap.

vascular Having a water-transport system.

Further information

WEBSITES TO VISIT

This website includes lots of useful information about plants and their classification, including some fun quizzes:
www.biology4kids.com/files/plants_main.html

Another useful website:
www.kidsbiology.com/biology_basics/five_kingdoms_life/classification1.php

This website includes some activities about classification:
www.fi.edu/tfi/units/life/classify/classify.html

This website explains how animals and plants are classified into groups:
www.zephyrus.co.uk/plantsoranimals.html

Index